EUROPE UNCOVERED

EUROPE UNCOVERED

SERAPHINA WILDE

CONTENTS

Introduction

The aim of this study is to shed light on common misconceptions related to geography and to geographically analyze Europe as part of this diagnostic process. With this goal, we have brought together 10 geography researchers who have experienced several geographical misconceptions and prejudices about Europe over and over again. In other words, the study tries to demonstrate the power of geography in dispelling groundless stereotyping using the experience of geography experts. We name the geographical misconceptions one by one and analyze them in this study. The research tries to 'uncover' Europe as the title suggests.

Methodologically, the study depends on subjective data and experience of the researchers rather than quantitative or qualitative data collected from others. We aimed to reveal the subjective experience of the researchers in the light of their choices of career. The paper is not defining Europe once again at the basic level, but scanning it through the eyes of the people who define, analyze, and teach it. The basic idea of the article is to demolish the stereotypes related to Europe considering its importance, function, and structure in the global scale.

Chapter 1: Europe's Size and Diversity

Europe is one of the most diverse places on Earth, and for its relatively small size, it has a high level of physical features. Within the current countries which make up Europe, every physical feature can be found except desert and tropical rainforest. However, Europe is much bigger than generally perceived. This is because of common misconceptions regarding the geographical limits of Europe, both east to west and north to south. The east-west spread of Europe was fixed by the Oder-Neisse line in 1945. There has been much confusion since about what countries can legitimately make a claim to be an integral part of Europe. Within modern geographic attitudes, this usually includes the countries of Armenia, Azerbaijan, the Caucasus, and Georgia, while other countries, including Russia, are considered only partially European.

Europe is a continent; a continent is a major landmass. It must have clear borders so that it is separated from the adjacent sea or land, and a number of countries must be present within the physical boundary of the continent. This definition has always been generally accepted, and this is confirmed by examination of the number of people that inhabited the European continent for at least the last two millennia. If the boundary were not fixed, the term European could have no meaning.

The Myth of Europe as a Small Continent

The myth of Europe as a small continent When we talk about Europe, we usually refer to it as a small continent, and this is especially true when we compare it to larger entities such as Asia, Africa, or America. This idea of Europe as being small coincides not only with cultural feelings and concepts inherited throughout the centuries, but this concept has a basis in common parlance. If one checks various urban legends and myths credited to Americans, one is quite likely to come across a joke about Europeans believing that all of Europe can fit in the state of Texas. The elements behind our assumption do appear to have substance. Before explaining the reasons, let's take a look at the hard facts. When we consider a list of the world's 30 largest countries, both by means of landmass and population, European nations are practically nonexistent with only six or seven appearing in any of these lists.

This lack is not that surprising when we compare the landmass of the European countries with the other 10 largest countries. With an area of 2.15 million km², France comes the closest, ranking 47th in the world. It is interesting to note that with a total area of Turkey of 783,562 km², and excluding Turkey, France is bigger than the largest of the Southern European countries. Portugal, in second place, is only larger than Mongolia, ranking 63rd in the world. Then we have Spain, ranked 52nd, and Sweden, ranked 55th. Only 10 ranks away, we find Germany, which is bigger than Nigeria but still below Turkey. The third of the Iberian Peninsula countries is Spain's immediate neighbor, Portugal, to be followed by the 48th ranked Poland. To finish, new data on geographic composition of countries' land area made available allows us, for the first time, to include the

Arctic's self-government bodies in the list, with Greenland close behind France, with an area of 2.166 million km^2, and the Faroe Islands similar in size to the Czech Republic.

Geographical Diversity Across Europe

Objective illustration of geographical diversity across Europe could immediately disprove many misconceptions. Though most believe in Europe's diversity, it is hard to comprehend the scale of this variation. Most parts of the continent could be considered relatively large relative to other parts of the world. This is because of the variety seen across its populous parts per se, irrespective of discernible cultural or environmental differences. The availability of data makes it easy to present singularly diverse regions. The illustration plotted below is difficult to categorize as deserts or tundra. Similarly, their hefty populations belie controversy of what constitutes continental interiors. The perky pattern of coastal or lower elevator loads is also information laden.

The unimaginable scaling-wise variations in Europe across festivities to the northwest, the resorts din to the Iberian head lodges or to the nirvana at Hispania peaks are once again very diverse if one recognizes variations within Spain itself. In addition, the Russian substantially regionalized cultural contracts are also uniquely peppered. Climatically, geomorphologically and agriculturally what is significant is that a latitudinal distribution very often cuts across almost all of the geography. It would be hard to develop a scale which could simultaneously cater to such a variety of zones so widely spread across a geographical divide on this minimal lineup. The conceptualizations may not be accurate enough to cater to all discernible criteria, but with enough data, requirements could be selected for enriching them typically for each set of experiences, enabling applications.

Chapter 2: Boundaries and Borders

Mountains and rivers, even languages, they are all traditionally offered as perfect natural boundaries. In creating stability, stopping invasions and providing identity. But, as we have seen above, these natural barriers can all exist in very different ways based on your perspective. And in reality they have rarely, if ever determined the edges of a land. Just picking pebbles off a beach tells us that natural boundaries are not stable. Rivers change their shape frequently through flooding their banks, causing loss of land at high waters while depositing fresh sediments at low levels. Mountains continue to rise and sink over thousands of years. Volcanoes can appear from nowhere and change entire landscapes, forever. Erosion and natural disasters can modify nature.

You may argue these changes are no more than evolutionary, few earthquakes are great enough to disturb established international borders significantly you may claim. However, imagine a country at a river's edge, according to law extending to the middle of the river. Due to modern agriculture, some sediment has been carried downstream and the land has eroded. An island now exists in the middle of the ex-riverbed. Who owns this island? The downstream country is going to need a powerful argument lest the upriver state decides to build a military installation upon it, or a holiday home for its president. Can anyone really build a stable boundary out of such mere pebbles?

Understanding Europe's Borders

Europe may not seem like a huge continent compared to Asia, and it may be easy to locate as one of the seven continents of the world. But with no obvious physical boundaries such as mountains, it becomes very hard for people to understand its limits and to come up with an adequate size. In fact, after the exploration of the New World, it was even easy to question the motives of how this single landmass with only one summit or edge, and in a north/south direction, could be a single continent and cover the broad range of different peoples and nations.

Europe can be considered to be a land bridge between the Asian and African continents. Taking a look at this image, only the Atlantic to the west and its islands, and the islands and similar projections and peninsulas of the Asian continent, stop Europe from being part of the single landmass that is Africa and Asia. Even then, the waterway around the European coast can be said to be systems of straits rather than the open sea.

The Ural Range and the geographic area between the Ural Range and the Volga River are considered to be the geologic border between the European and Asian continents. Similarly, the Kuma-Manych Depression is the traditional border between Europe and Africa. The Don and the Water-Manych, a tributary of the Don, send their water to the Black Sea and the Caspian Sea basins respectively. For Russia, the border between Europe and Asia is actually calculated depending on which drainage system a river is in, as the drainage divide is usually taken as the border. However, the standard flaw in this continent delimitation is that the Ural Range does not act as a mountain barrier in the way that the Alps or the Himalayas do. Europe does not end at the Volga River, and much of this river is in Europe. Islands in the river, particularly at elevations, make it possible to cross the Ural Range without ever leaving Europe.

The Concept of the European Union

The European Union (EU) is a unique economic and political partnership between 27 European countries. It is a political and economic union of 28 member states, located primarily in Europe. This began a process of convergence that towards the end of the century would have grown into the European Economic Community (EEC) and the European Atomic Energy Community. The EU unifies member countries to achieve objectives no single country could reach alone. It is an economic union - a single market and customs union with its goal the establishment of the euro currency. In the past, the EU's goals included raising living standards in member countries, maintaining peace between them through an intergovernmental system of cooperation and the maintenance of a highly developed single market. The EU represents the largest situation of independent well-being worldwide.

The European Union is based on the rule of law, whereas every action taken by the member states and the member organizations can be questioned in courts. The European Union decided to protect workers and thus designated a specific government, with an independent central bank and controlled by the interminable monitoring capacity of European Union institutions, as a special government area, with specific rights and goals of operation. In a few years, from nine European countries, a group of countries have formed the greatest geo-economic aggregation on the planet: the European Union. In no continent have common political, military, and economic institutions been established. Among European countries, no such principles have been established to ensure that differences will be resolved without resorting to war concerning any situation involving any group of interest. The European nations, losers in the strategic phase, tried to defend themselves against the economic superiority of America and Asia with a formal transformation of the agreements.

Chapter 3: Mountain Ranges and Rivers

Mountain ranges and rivers - Europe Uncovered: Debunking Common Geographical Misconceptions

When you think about the largest mountains in Europe, the Alps come to mind. However, Europe has a vast range of immense mountain ranges. According to the Alpine Convention, ratified by the Alpine range countries and the European Union, the Alps cover an area of 190,956 km² stretching over 1,200 km at an average width of 100 kilometers. The highest point is 4,810 meters, reached at the summit of Mont Blanc.

Eurasia contains a massive area stretching from the Atlantic to the Strait of Malacca. The peninsula has mountains in a number of surprising locations. The Crimean mountains (the highest point is Roman-Kosh, 1,545 m), Sudetes Mountains (the highest point is Śnieżka, 1,602 m), Ore Mountain (municipal unit) (the highest point is Klínovec (Keilberg), 1,244 m), and Rila–Rhodope massif with the mountain Mussala (the highest peak on the Balkan Peninsula, and in the European territory of the former Soviet Union, with an altitude of 2,925 meters). Europe has two giant mountain ranges extending from west to east. The Alps in Europe and the Ural in the south are the smallest – the Anonymic. The highest peak of the Ural, Narodnaya, has an altitude of 1,895 meters, making it one of only a few mountains in European Russia over 1,800 meters high. The highest peak of the Alps in Europe is located at the 2,301.90 meters-high Mont Blanc Mountain in the Alps. It is considered the largest mountain in European territory.

The Alps: More Than Just Switzerland

When you think of the Alps, it is easy to think of only Switzerland, but that is not the only country that holds what seems like an infinite list of cool sites. The towering peaks of the Alps stretch across a sizeable chunk of Europe that includes eight other countries including France, Italy, Germany, Austria, Slovenia, and the smallest nation in the world, Liechtenstein. Yes, the Alps actually cover almost 300 miles, the same as the drive between New York City and Richmond, Virginia. Since it is spread across so many European nations, the Alps are home to tons of indigenous animals that you cannot find anywhere else in the world. Some of these animals include the ibex, the alpine marmot, and the elusive lynx. There are also over 7,000 species of plants that have been counted in the European Alps. There is also a good bit of lakes located in the mountains, the largest ones being Lake Geneva and Lake Constance.

Helping cut Switzerland in half is the tallest mountain range in Switzerland, the Pennine Alps. This mountain range is where you can find the Matterhorn. With 22 peaks to summit, who knows you might even see the Abominable Snowman. The Pennine Alps are not only in Switzerland, it stretches out into Italy as well. It is on the Italian side that you find the highest peak of the mountain range, Mont Blanc, at almost 16,000 feet. Also at these heights, you do not find any treeline or vegetation, so the land is called an alpine climate. So, the next time you start thinking of all things Switzerland when someone says the Alps, you might want to check the other eight countries that the mountains are connected to.

The Danube: Europe's Second-Longest River

The inhabitants of the countries through which the river passes regard the Danube as an important element spicing up their local landscapes, national economies, and cultural spectrum. To a certain degree, there is even rivalry between the countries as to which section of the river is the most attractive, cleanest, and most interesting. Austrians speak of "Europe's most beautiful stretch" when referring to the Wachau Valley, Slovaks like to call their Komárno-Bratislava section "The Queen of European rivers," while Romanians vie for bragging rights over the portion from the Iron Gates to the Danube Delta. In the world having approximately 2900 international rivers, surely one, however, must be the second-longest? This role is ascribed to the Danube River, the longest river in Europe, and 4th longest in the Continent. The river passes through more countries than any other in the world, traveling a mere 90 km in Moldova, and twice as many in Croatia or west of Romania's capital. Even though the Danube is much shorter than other great world rivers, it easily outstrips them regarding its speed and flow volume at the mouth.

The 2857 km crossing Europe are home to at least 23 nations and have given birth to Europe's biggest towns and the majority of European nations. Thirty towns located on Europe's riverside are over 100,000 inhabitants, with 17 of those firing at least 200,000. Along centuries, the inhabitants of the continent have built hordes of dams, locks, and canals, many of which can take pride in being unique tourist attractions. This extensive network offers unforgettable cruising opportunities, hundreds of nations being lured in every year by the magical aura of a once unknown, powerful river. The secret of the Danube River's enormous intrinsic force, of it drawing people's attention to itself as easily as a double-decker bus would, is nothing new. Throughout its journey, the Danube has always been about blooming, chaos, and decay, about trade and shipbuilding, about the encounters between different cultures, and about the competition between empires.

The river and its tributaries have knocked down the ancient Limes roads that traversed Europe, and through its "right of asylum" policy, has provided a safe environment for displaced people. The many billiard games the European empires played for its possession provide ample evidence of how valuable the Danube River has always been. Cruise ships are proof that the river has lost none of its charm. The residents of the nobly adorned riverside towns gather on the docks to catch a glimpse of those lucky enough to spend most of their charmed lives cruising away day after day into the roaring tranquility of the Danube River.

Even on the most peaceful countryside stretch, the river is nothing but a great show of splendor. From the White Forest to the Black Sea, from great shipbuilding centers to the enchanted Danube Delta, from the concrete gates to the exciting suspension bridges of Vienna, Budapest, Bratislava, and Belgrade, the alarm lights of the river tell remarkable stories. Their countries have to confess the Danube River is one of the few things that unite them not to mention they are constantly urged to rethink their relationships by those 7 to 28 cubic kilometers of water flowing into the Black Sea. If the Danube was ever given a conscience, it would constantly question its age: it does not want to be kept alive by history only but to continue sowing the seeds of future diversity for those it serves in its magnificent role of ancestor.

Chapter 4: Islands and Archipelagos

The European continent has always been open and exposed to numerous influences that derive from relations present in one way or another, or from spaces that, without having the physical condition of insularity, assume these characteristics indistinguishable today. Island behaviors, especially in those ages when navigations and transports are very slow, cannot count on today's large and small openings, which are only partially available in today's vehicles, have little contact with the current traffic tides plaza exposure.

Yet, incredible behavioral passages are born here, especially when the islands are full of natural resources. Strong connections and availability that the current degree of technology would suggest an opposite reality. But imagine going back to the transoceanic navigations and, just at the end of the island, an outlet of immense importance to interact with the travelers and the needs to continue the journey, leave to increase to an unthinkable value the problem of lack of opening. We must also assume that some doubts have reason to exist, also because some islands of immense success, despite relative flatness, have real communication membranes, many of which in all respects can offer extensive relations with the outside world.

Exploring the Greek Islands

Greece, particularly its islands, fascinates and captivates us due to the cultural and strategic events in which it has played an important part. Equally fascinating is the strategic and environmental importance of the Greek islands. Yet, besides a few known examples, they seem like secret portholes into a surreal other world, tradition-filled corners where the olive tree reigns supreme and the sound of 'Zorba's dance' softly accompanies the tender drama of warm, sea foamy waves lapping the shore. Some Greek islands vibrate and also echo to the cries of 'Sirtaki', the celebratory Diogenean outburst that draws tourists to the Federation of Dodecanese islands, namely the ring of twelve bastions that defend part of the Aegean from the Southwest of Greece up to the coasts of Icaria, Chios and Samos to the North of Asia's Minor.

The Greek islands – more than 3,000, 169 are inhabited, administration is mainly provincial, tourism, fishing and agriculture are the main activities – represent an expression of particular concepts and values, ranging from opposing destructive mass-tourism to responsible tourism, blending tradition and hospitality with the need for a solid equilibrium between admiring natural environments and consideration of the right to access... where the fascinating Greek way of life is different from ours – Islands that are sea and mountain, even steep cliffs.

The British Isles: A Complex Geographical Entity

It also highlights the complexity of referring to a particular place by giving it a name and assuming that others understand which precise place you mean. There are many places elsewhere in the world that share this complexity but it is particularly striking in the case of the British Isles. Great Britain is the largest island in the British Isles and it lies to the north-west of mainland Europe. In Great Britain, the main constituents are Wales and Scotland to the west and north, and England to the east of a line linking the mouth of the River Dee with the Trent and the Humber.

Ireland is the second-largest island and lies to the west of Great Britain. The island of Ireland is politically divided into Northern Ireland (part of the United Kingdom) and the Republic of Ireland (an independent state), and although this is convenient as far as administrative practices are concerned, it is rather more complex when it comes to geographical description. Just what should we call the group? This is not an easy question to answer, and it is important to be aware of the identity of the group that is being covered by the term. As we shall see, some of the problem is caused by the fact that the name is political rather than geographical, and although nominally referring to islands, its use extends in varying degrees across the two largest islands but not to all the smaller islands in the group.

Chapter 5: Climate and Weather Patterns

Chapter summary: You have probably heard about Scandinavia, the place known for its freezing cold. While parts of Scandinavia indeed can be freezing, the cold is usually just skin-deep. The climate is affected by a warm sea current originating in the Gulf of Mexico. Bad weather over the British Isles is often referred to as typical British weather. A fact not so many of us reflect upon is that 'typical British weather' is quite similar to the weather over the Norwegian coast, west of the British Isles. Both places are affected by warm sea currents. While spring usually starts much earlier and fall usually arrives much later in Europe than in the central parts of the USA, summer temperatures can rise high above 40 °C in the Mediterranean. Warm air, dry air, and direct sunlight help build up the heat. Yes, direct sunlight is not as easy to take while standing at the South Pole. The sun's sunbeams are less likely to hit you than in southern Italy. The Long Term Daily Satellite records global cloudiness. The clouds usually gather where we least need them - at home, in Europe. Scandinavia's cold can be skin-deep. In every way.

Weather and climate along the world's longest coastline - northern and fresh air. Those of us interested in history sometimes discuss what historical leaders really have in common with the general population. How many leaders could really share the problems of cold, disease, starvation, or fears for the weather? Well, according to some historical sources, the slaves (or helots in old Greek) were not allowed to hold very high commandments. The helots did all the work while the Spartans enjoyed all the privileges. Military decisions were made while enjoying clean air that you or I can feel through our lungs every single morning we spend at our summerhouse in the Swedish archipelago during the summer. The slaves had never seen a real coast. Never experienced the real chilly northern wind.

A common saying is that history repeats itself. If weather patterns remain constant enough, this story should be reliable. The history of our entire continent has probably started over a morning coffee at sunrise, at a coast in the south. First when trading increased, and farmers understood the possibilities of using salt to conserve food, were the other people of our old continent given the possibility to live at least a part of their lifetime at a coast as well. The colder you move north, the fewer people live at the coast. And still, history has taken the same path. Today, most of the world lives at the coast or in coastal zones. This might sensibly be explained due to such a high influx of immigrants. Half the world's population has immigrated to such zones. If slime was so rich in Sweden, could we really believe we were living in the wealthiest of all countries?

The Mediterranean Climate Zone

Introduction: The Mediterranean Basin contains some of the world's most dramatically changing terrain. Surrounded as it is by the Atlas, Pyrenees, Alps, Alps-Volcanoes, Dinaric-Balkania and the Hellenic-Hercynian Ranges, sudden and remarkable changes are experienced by both airplanes and travelers unaware of the speed of their passage and of the vast structural and climatic differences of the land beneath them. In dramatic contrast is the generally very hot climate of this region in the great area entirely enclosed by the ranges mentioned above. Only the frequent winds and the notable exceptions of a summer sea breeze keep this region from being as uninhabitable as the Sahara in Africa to the South, the warm deserts to the East and the hot scrub country of Spain's Southern Meseta to the North. It is a climate zone unto itself and has been named the Mediterranean type climate, a term that is too often confused with the term "hot and dry" climate.

Definition: The Mediterranean Zone is characterized by a lack of or slight winter rainfall, or a slightly wetter spring, and frequently by a paralyzing dry season in the late summer and autumn. Economic plants such as olives, figs, and grapevines that have fine root systems are watered by torrential rains in spring and winter, local artesian water and have been domesticated by Man in a sub-soil environment that is extraordinarily moist. These plants have evolved badly developed protective devices for their leaves and are highly flammable under hot and dry conditions. For many years, Man has spent great effort and not a small amount of ingenuity destroying his built-in allies that are his rulings against fire. The Zone type was named from the Mediterranean Basin of Southern Europe, as most plainly exemplified by the State of California on the North American Continent. The spatial arrangement and makeup of the smaller geographic regional units are derivative of the progressively shifting position of the Intertropical Convergence Zone that more effectively has its mid and late summer precipitation silenced by the Mediterranean Complex.

The Influence of the Gulf Stream
We have received various questions concerning the temperatures in different regions of Europe, especially places in countries thought

to be cold, like Norway or Russia. A significant influence on the climate in Western Europe is the Gulf Stream bringing warm water from the tropics. The warm water heats the air and circulates over Western Europe. For example, Paris has the same location as Manitoba, Canada, which has its temperature strongly influenced by the cold arctic air. However, Paris has the warm air from the subtropical climate and, as a result, has a much milder climate despite having the same latitude as Manitoba. Also, the water beside Europe saves the continents from extreme temperature variations. While Kiev has cold winters and hot summers, mainly influenced by the distance to the sea, Astana and Chelyabinsk have much colder winters than Kiev.

However, the water also explains the European climate dissimilarity. Popular political scientific studies rejecting biogeographical facts due to historical causes addressed Europe. When the Dutch would have lived in the south and the Italians in the cold climate area of Europe, history would have turned out differently. However, climate facts speak against these cultural and genetic explanations; more than against the homeland hypothesis. Italy and Southern Spain, one of the most historical countries in European history, have their territory located in an area of subtropical weather – Southern Europe. In this scenario, the drinking habits also could have turned out differently. Wine being a typical Southern Europeans' drinking, a result of leaving the grapes in the sun, and beer emerging from the North due to a lack of long sunshine periods.

Chapter 6: Urban vs. Rural Dynamics

Urban vs. rural is the oldest resilience debate in at least the Western world, where the city has for centuries epitomized decadence. Decadence, in turn, is the antithesis of resilience. But that debate was always narrow, and shades have also had their say: abundance/privilege creating not only fragility but innovation, invention, and interconnectedness. This urban/rural antagonism can be observed from the earliest notions of perceived 'European identity' described by Herodotus and Aristotle, to the original biblical dichotomy of the city of Babylon and the suburban Garden of Eden.

Sidney's Arcadia developed as an archetype of a country's virtues, but only possible after some considerable wealth and education had been created elsewhere in the city. The Head Man versus de Saussure both observed intolerable rural lives a century apart. Urban life was 'on a pedestal,' as romantic tales, just as Hermann Hesse's Siddhartha, reinterpreted recently by Paulo Coelho in The Alchemist. Nor is it a dichotomy purely from the classic Western culture. Little has changed in tropical rainforest ecological terms; the roots of the banyan tree rapidly span the space where others shed leaves to resume the unending battle against decay necessary to support the diversity of interlopers gathered around its predictable stability. Moscovici's study of how a creative minority is needed for societies to be 'dense and varied' shows that cities have always been the most significant generators of this creative minority. Such dissidence is an indicator of richness, and harmony or consonance (or as one philosopher put it 'a level of coincidence consistent with continued tolerance of the energy exchanged between opposites') important constituents contributing to resilience.

Megacities vs. Small Villages

When thinking about Europe, one likely concentration is that it is a densely populated area with urban centers, mega-cities such as London, Paris, Berlin. Yet like any other continent in the world, Europe features many small towns and villages that are dotted across the landscape and have retained their character and lifestyle for centuries. Europe forms an area which has a strong culture based on long-standing historical heritage. There are many visible ancient monuments spread into the small streets of the beautiful old towns. Hospitable people are still following the traditions and customs that have been guiding their lives for many previous generations.

This is the Europe that we all enjoy and love, but unfortunately there are also some shortcomings that need to be addressed in the modern digital age. The physical position of small towns and villages – remote, distantly separated, and requiring long travel times – is also reflecting a certain distance in terms of modern communications and technology. This is something that the region is aware of and is actively seeking ways to address in order to offer an interesting lifestyle to live in, and therefore increasing its attractiveness for the citizens and other individuals who are seeking new opportunities.

Urbanization Trends in Europe

G. Kohlheb. Globalisation and Enlargement: The Nature of the Challenge Facing Territorial Diversity and Sustainable Development in Europe. Part II: The Concentration of Persons and Economic Activities in Certain Urban Areas in Europe. Chapter VI. Concentration of Population and Employment. II. Urbanization Trends in Europe. August 5th, 2002.

In the last 50 years, the phenomenon of mass urbanization, defined as the concentration of a large portion of a country's population in cities, which is also accompanied by a decrease of the population in "rural" areas, has taken place in most developed countries. Whereas 75 years ago, 70% of the population in the United States lived in the countryside, today, only 20% do so. As a result, the concentration of population and economic activities in the large urban areas is augmenting, and "cities" with a population of several million are arising, such as in the United States where 19% of the total population lives in 15 major urban regions of more than 5 million inhabitants each.

Similar urban centres are also visible, although not as many ones exceeding a population of 5 million, in several other developed and emerging economies, such as in Japan, or the Sao Paulo and Rio de Janeiro regions in Brazil. The question arises, how do European countries compare with urbanization patterns elsewhere, and in terms of national data, and across the EU, what are the main general inhabitants mobility and spatial concentration trends liable to cause in order to raise the territorial cohesion challenge? Within Europe, the "urban belt" region prone to the highest urban densities stretches from southern England to northern Italy. The Randstad Region in the Netherlands, the Rhein/Ruhr area in Germany, northern Italy and Belgium belong to the heavily industrialized regions of Europe, and suffer from the effects of agglomeration. Due to these regional concentrations of population, the Netherlands, Italy and Belgium are some of the largest functionality-oriented units, which perform both economic and structural tasks.

Chapter 7: Transportation and Connectivity

Transportation Factors A. Large Rivers: Several significant rivers pass through the country, aiding the transport of goods and people. B. Extensive Canal Network: Some countries also feature extensive canal networks, with the Netherlands being a prominent example. C. Location on the Global Trade Route: Countries on the Mediterranean Sea benefit as they are key links between African, Asian, and European trade routes. D. Coastal Ports: Europe also has many large coastal ports, along the Atlantic and Mediterranean. The ports are important since they are the final destination or starting point of goods transported on ships. E. Importance of Air Transport: The country should be connected to major aviation networks since most of these are cargo flights.

Cultural Factors A. Recognition as a European City: In addition to tourism purposes, people do not recognize Asia as being part of Europe. B. The Importance of Location: Location has significant impacts on the cost and time of shipping goods, so companies often prefer setting up shop in a major European city.

The Importance of Rail Networks

While the competition to attract increased air traffic is fierce and major airports have facilities to pamper high-yielding international business travelers with their Platinum, Silver & Gold Status cards, the "domestic market" for air travel is at least as important: thanks to railway operators being allowed to compete on an equal footing with the bus and motorway network operating interregional services, around 75% of air journeys that can be made in under three hours by a train are now undertaken this way. In Germany, and between Zürich and Milan, the figure is as high as 90%. In the case of connectional flights that are shown as stopovers in the airline timetable so that the established carriers can demonstrate their worldwide presence, 80% of them are in fact rail journeys made on so-called feeder routes. In France, the TGV line connecting Paris and Charles de Gaulle airport is the most profitable section of the entire TGV network.

This great demand is due to fast, attractive services; thanks to being able to compete on domestic routes without the constraints that are imposed on them on international routes where bilateral constraints are in place, and the economic plight of many national carriers, railways often have a superior product. The check-ins for air services require passengers to arrive at the terminal early, and anywhere between a 20 and 45-minute ride to most European airports, be they served by a chauffeured limousine, bus, or an undeveloped, congested branch line. Many train services have downtown departure and arrival points offering city center to city center transit times that would put the LCCs to shame and similar class two airports: fares can be from 33% to 65% cheaper. This wide range depends on the type of service on offer, and the travel class and esertia for the time of booking.

Air Travel in Europe

It's easy to misjudge the size of your next European vacation. The most common misconception is that it's a quick hop to most places. In most countries, even with the high-speed trains, it takes at least a couple of hours to travel to a major city. Traveling in northernmost Norway is more comparable to Canada than Lofoten. As for London, don't think it's too far-fetched to spend more time traveling to Cornwall than Paris. While driving times are helpful in planning, they underestimate how much time one should spend traveling in parts of Europe. After all, people do travel, visit, and live in the farther out regions. Though it's not smart to skimp on travel time upon arrival, either. Something seems to be lost in translation because Europeans love another form of travel and it'll take a bit of time, a lot of strategizing, and patience.

Within Europe, the biggest business and vacation distance myth is that everyone has access to abundant low-cost flying, let alone flying within everyone's country or region. There are two reasons airlines would not operate internal flights nearby: they are likely too short, crowding airways intended for longer flights. So. If you want a package vacation in the coziest knights and little band of history or prefer the dramatic coastline of Pembrokeshire, start thinking how the tales of Chaucer begin conquering a fear of the railway or the seat of the quick cross-channel ferry. However, it's very likely the mainland railway or freeway routes in most countries can effectively transfer you to further destinations for the same time and price as flying. Indeed, save precious time and connection times if you're challenged with going to Paris as a hub event.

CHAPTER 9

Chapter 8: Geopolitical Realities

While most often residing under a geographic heading or puzzle, the makeup of Europe is also the expression of an early geopolitical competition that shaped the state system in its interior in the course of the 17th century. In the process, its northern parts became more nation-state-like and have continued expanding nation-state principles through the establishment of regional governmental bodies and the creation of 'acquis communautaire'. Europe's geopolitical realities have become the playing field for citizen-centered societal models called democracies, with the northern part resembling the dualism of Britain more than any other European monarchy and the southern part following the emergence of complex cities and medieval universities. Geographical limitations have channeled the progression of nations differently, echoing regional differences in a myriad of fields.

Political formation of such a diverse set of states was regulated by Europe's shape, just as politics could contour its shorelines in a spiral movement of expansion. The political competition for control of Europe's varied agricultural and mineral resources transformed poor, egalitarian territories into territorial states that would first breach Europe's fortress in 1492 and challenge and conquer large world regions over the two centuries following. Security needs gradually shaped the cultures of these pioneering states, putting a militarized and power-driven distribution of sovereignty into a hitherto barely organized common administrative area. Europe has now defined a governance model that negotiates and shares rules through a mix of agreement, anarchy, and submission. European states move back and forth between national, provincial, and local commitments, and their tenure over a specific territory can end following demographic cycles of settlement, conquest, departure, and return that solidify state power stakes.

The Balkans: A Region of Historical Conflict

The Balkans has a long-standing reputation for being a region of conflict, and as a result, it carries a lot of baggage in the eyes of people outside Europe and its local residents. There is still historically rooted animosity between people from different countries in the region, with interethnic relations, dependence on interfering external powers, and the repercussions of opportunistic politics playing a role. One part of history that people find hard to let go is that the region does not have a homogeneous ethnic composition, but there are twenty-nine fiercely tells the story of how different groups fought over a power vacuum created by the dissolution of Yugoslavia until about the end of the 1990s. This period marked a time of massive internal and external interference, including a soporific collaboration between the European Union and Russia.

While there is a historical angle to the infamous Balkan stereotype, it is quite a narrow perspective. The wider Balkan region holds great charm for tourists and, to this day, remains relatively undiscovered with villages pretty much untouched by the industrial might of central Europe. It would be nice to believe that people's views of particular regions of Europe should be based on what is there now rather than rough patches of time, with negative views being held for centuries to come. Hopefully, enough associations between Viennese waltzes and Balkan cheap thrills and Dubrovnik and Sarajevo for Westerners. Topics such as Bosnia's Hausa minority and the recent gay festivals in Albania remain a bit too obscure.

The Impact of Colonialism and Imperialism

Colonialism could indeed explain why place-names have been changed or why many cities have less "indigenous" population than one could expect. Within most national borders in the Americas, there was once a distinct indigenous population. Today, that indigenous population has vanished, been killed by disease, dragged away in slavery, or assimilated with the growth of other ethnic groups. The "few" remaining people are suppressed due to racism, guilt, or the association with the least successful people of the country. From the moment the first Europeans started exploring the world and came in contact with other civilizations, the world's history was illuminated by their presence.

Large empires, foreign colonies, imperial aspirations - after more than 500 years of relations, normal relations are still a rarity. The people living in the colonies were descendants of various wealthy civilizations. At their peak, the Aztecs ruled over a densely inhabited empire with large cities and extensive irrigation systems. Their empire had a complex social and religious structure. At this time, cities in Europe were of little importance, except maybe Rome and Paris, which could accommodate a few thousand people, or Beijing, where over one million people lived.

CHAPTER 10

Conclusion

Myths of the world are quickly dispelled when we actually take a moment to critically think about them. Cases show that facts are often preferred above myth, stereotyping or prejudices. However, critical thinking or fact-checking cannot guarantee always perfect disguising of one's personal identity. In the first part of our article, the emphasis was put on our urban mindset, so that we take perspective of the dwellers of the countryside in Europe.

We compiled our maps and tables and added some subjective commentary. The material is aimed at showing new perspectives of Europe, in order to make European rural life a more familiar and recognizable for an urban reader. The present work is 'in progress' meaning the compilation will be extended and improved. Our aim is to let the readers share the view we have, and together to discover Europe through the lens of its countryside. New results and viewpoints will be used in classes on rural development and spatial analysis. Tables and maps have been compiled with free, open source GIS and statistics software, making it possible to build an open-access educational atlas on this material. Feel free to share your thoughts on this work.

CHAPTER 11

References

Ad, H., Van, J. C. P., & Kloesik, R. (1996). Environmental Systems and Societal Development: ISDP Project Handbook. Kluwer Academic Publishers, Dordrecht, The Netherlands. Cinnirella, S., & Murgatroyd, P. (2018). Education and the Brexit vote: why did some young people reject contemporary mainstream modes of political engagement? Political Studies, 66(2), 508-525. COM/2EN/3. Daw, T., & Gray, T. (2005). Fisheries science and sustainability in international policy: apprehension, confusion and omission. ICES Journal of Marine Science, 62(6), 1094-1104. European Commission. (2012). Risk on, risk off? Portfolio allocation of European consumers, consequences for their foreign exchange exposure and the impact on emerging economies. Consumer Conditions Scoreboard.

Eurostat (2008). Program for the Modernization of European Enterprise and Trade Statistic (MEETS). Prades, L., & Adriaan, P. (2003). Estructura espacial y patrones de localizacion de las Sociedades Anonimas en Espana. Investigaciones Regionales, 2(3-4), 77-97. UNECE (2007) Definition of core, developed and other countries. in UNECE, Trade and the Digital Divide. 309. Word Trade Organization. Yu, S., & Mol, A. P. (2018). The dynamics of environmental non-compliance within East Asia. Journal of Environmental Management, 216, 158-170.